Oregon
Total Eclipse Guide

Commemorative Official Keepsake Guidebook

2017 Total Eclipse State Guide Series

Aaron Linsdau

Sastrugi Press

Jackson Hole

Copyright © 2017 by Aaron Linsdau

All rights reserved. No part of this book may be reproduced or transmitted in any form or by any means, electronic or mechanical, including photocopying, recording, or by any computer system without the written permission of the author, except where permitted by law.

Sastrugi Press / Published by arrangement with the author

Oregon Total Eclipse Guide: Commemorative Official Keepsake Guidebook

    The author has made every effort to accurately describe the locations contained in this work. Travel to some locations in this book is hazardous. The publisher has no control over and does not assume any responsibility for author or third-party websites or their content describing these locations, how to travel there, nor how to do it safely. Refer to official forest and national park regulations.

    Any person exploring these locations is personally responsible for checking local conditions prior to departure. You are responsible for your own actions and decisions. The information contained in this work is based solely on the author's research at the time of publication and may not be accurate in the future. Neither the publisher nor the author assumes any liability for anyone climbing, exploring, visiting, or traveling to the locations described in this work. Climbing is dangerous by its nature. Any person engaging in mountain climbing is responsible for learning the proper techniques. The reader assumes all risks and accepts full responsibility for injuries, including death.
Park maps are courtesy of the National Park Service.

Sastrugi Press
PO Box 1297, Jackson, WY 83001, United States
www.sastrugipress.com
Quantity sales: Special discounts are available on quantity purchases by corporations, associations, and others. For details, contact the publisher at the address above.

Library of Congress Catalog-in-Publication Data
Library of Congress Control Number: 2017905381
Linsdau, Aaron
Oregon Total Eclipse Guide / Aaron Linsdau-1st United States edition
p. cm.
1. Nature 2. Astronomy 3. Travel 4. Photography
Summary: Learn everything you need to know about viewing, experiencing, and photographing the total eclipse in Oregon on August 21, 2017.

ISBN-13: 978-1-944986-07-0
ISBN-10: 1-944986-07-3

508.4—dc23

Printed in the United States of America
All photography, maps and artwork by the author, except as noted.

10 9 8 7 6 5 4 3 2 1

Contents

Introduction	4
All About Oregon	**6**
Overview of Oregon	6
Weather	11
Forest Fires	12
Road Closures Due to Fires	13
Wilderness and Forest Park Safety	14
Eclipse Day Safety	17
All About Eclipses	**19**
Total vs Partial Eclipse	20
Early Myth & Astronomy	22
Contemporary American Solar Phenomena	24
Future American Eclipses	26
Viewing and Photographing the Eclipse	**27**
Planning Ahead	28
Understanding Sun Position	29
Eclipse Data for Selected Oregon Locations	31
Eclipse Photography	**31**
Eclipse Photography Gear	34
Camera Phones	36
Viewing Locations Around Oregon	**43**
Remember the Oregon Total Eclipse	**65**

Introduction

Thank you for purchasing this book. It has everything you need to know about the total eclipse in Oregon on August 21, 2017.

A total eclipse passing across the United States is a rare event. The last US total eclipse was in 1979. It traveled over Washington, Oregon, Montana, and the corner of North Dakota.

The next total eclipse over the US will not be until April 8, 2024. It will pass over Texas, the Midwest, and on to Maine. After that, the next coast-to-coast total eclipse will be in 2045!

It's imperative to make travel plans today. You will be amazed at the number of people swarming to the total eclipse path. Some might say watching a partial versus a total eclipse is a similar experience. It's not.

This book is written for Oregon visitors and anyone else viewing the eclipse. You will find general planning, viewing, and photography information inside. Should you travel to the eclipse path in Oregon in mid-August, be prepared for an epic trip. The state estimates a half million visitors will converge on Oregon.

Most hotels in the communities and cities along the path of totality in Oregon have already sold out as of the writing of this guide. Finding lodging along the eclipse path will be a major challenge.

Resources will be stretched far beyond the normal limits. Think gas lines from the late 1970s. It may be likely that traffic along mountain highways will come to a complete standstill during this event. Be prepared with backup supplies.

Many smaller Oregon towns are far from any major city. Oregon mountain roads are slow. Please obey posted speed limits within all forest and park areas. Be cautious about believing a map application's estimate of travel time in Oregon.

People in all communities along the path of the total eclipse plan to rent out their properties for this event. With a major celestial event in the summer of 2017, be assured that Oregon "hasn't seen anything yet."

Is this to say to avoid Oregon or other areas during the eclipse? Not at all! This guidebook provides ideas for interesting, alternative, and memorable locations to see the eclipse. It will be too late to rush to a better spot once the eclipse begins. Law enforcement will be out to help drivers reconsider speeding.

Please be patient and careful. Large, beautiful, but dangerous animals make the forests in Oregon their home. Throughout the year, hundreds of animals are killed on mountain roads. Black bear, deer, and coyote are difficult to see along the highways.

You should feel compelled to play hooky on August 21. Ask for the day off. Take your kids out of school. They'll likely be adults before the next chance to see a total eclipse. Create family memories that will last a lifetime. Sastrugi Press does not normally advocate skipping school or work. Make an exception because this is too big an event to miss.

Wherever you plan to be along the total eclipse path, leave early and remember your eclipse glasses. People from all around the planet will converge on Oregon. Be good to your fellow humans and be safe. We all want to enjoy this spectacular show.

Visit www.sastrugipress.com/eclipse for the latest updates for this state eclipse book series.

Author Information

Polar explorer and motivational speaker Aaron Linsdau's first book, *Antarctic Tears*, is an emotional journey into the heart of Antarctica. This Jackson Hole native ate two sticks of butter every day to survive. Aaron coughed up blood early in the expedition and struggled through equipment failures. Despite the endless difficulties, he set a world record for surviving the longest solo expedition to the South Pole.

Aaron teaches audiences how the common person can achieve uncommon results. He shares his techniques for overcoming adrenaline burnout and constant overload. He inspires audiences to face their challenges with a new perspective. Aaron builds grit, teaches courage, and shows how to maintain a positive attitude in the face of adversity. He hopes that you will have a comfortable and enjoyable time watching the total eclipse in Oregon.

Visit his websites: www.aaronlinsdau.com and www.ncexped.com

All About Oregon

Overview of Oregon

Oregon, the Beaver State, is one of two states in the region referred to as the Pacific Northwest. Portland is famously rainy. Though gray skies are plentiful, that also makes the city and nearby forests incredibly green.

Located on the coast of the Pacific Ocean, Oregon is home to some of the most beautiful beaches in the world. The beaches swarm with thousands of visitors in the summer from the cities along Interstate 5. Both tourists and locals enjoy swimming, beach volleyball, sunbathing, and water sports. Children build sand castles and join their parents in relaxing along this scenic coastline.

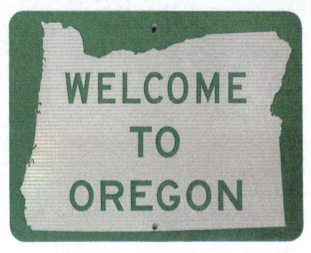

Although the beaches are beautiful, the Oregon coastline can be dangerous for the inattentive. Logs, rocks, and "sneaker waves" have all taken lives. If you plan to visit the beach, please read and heed the warning signs at beach parking lots.

Oregon is one of the many states in the United States that boasts numerous water bodies and unexploited water sources. The state is home to one of America's most famous rivers, the Columbia River that is situated on the northern border with Washington. While touring Oregon, make sure to visit the Columbia River Gorge National Scenic Area. Follow the river as it cuts through the Cascade Mountain Range. The river forms spectacular views all along its pathway.

Oregon expects hundreds of thousands of visitors for the total eclipse on August 21. The small city of Madras expects massive numbers of people, ballooning the population from 6,000 to over 100,000 over the course of the total eclipse weekend.

Hotels and Motels During the Eclipse

Portland has experienced a continual population increase since 1990. The movement of Americans into large urban settings has put pressure on many cities along the Interstate 5 corridor.

Once word of the total eclipse over Oregon spread, rooms became scarce. Nearly every hotel in towns along the path of totality has been sold out for a year or more. Oregon is not alone in this challenge. Hotels in Idaho and Wyoming have been sold out for multiple years in anticipation of the total eclipse.

What does this mean for eclipse visitors? Lodging and room rentals in eclipse towns will be at a massive premium. Does that mean all hope is lost to find a place to stay? Not at all. But you will have to be creative. There will be few if any hotel rooms available in these eclipse cities by the time this book is printed. Accommodations in the cities and towns along the path of the eclipse have been sold out for months.

In spring 2017, the author searched on Hotels.com for rooms along the total eclipse path on the weekend of August 21 and found many cities and towns along the eclipse path were completely sold out. This includes Salem, Oregon's capital.

Search for rooms farther away from the eclipse path. If you are willing to stay in Portland and drive to Salem, rooms were available in early 2017. As the eclipse approaches, people will book rooms farther from the totality path. By midsummer, rooms in cities like Portland may be unavailable. The effect of this event will be felt across the country.

Tour buses routinely drive for hours from Portland to visit beach communities along US 101. Think regionally when looking for rooms. Be prepared to search far and wide during this major event. If a five-hour drive is manageable, your lodging options greatly expand, but it also increases your travel risk.

Internet Rentals

To find rooms to stay in towns along the eclipse path, try a web service such as Airbnb.com. Note that many people rent out rooms or homes illegally, against zoning regulations. Many cities have already begun to feel the crunch of eclipse inquiries.

If Oregon towns fully enforce zoning laws, authorities may prevent your weekend home rental. Online home rentals during the eclipse will be a target for rental scams. People from out of the area steal photos and descriptions, then post the home for rent. You send your check or wire money to a "rental agent" then show up to find you have been scammed. If the deal sounds strange or too good to be true, run away.

Camping

If you can book a campsite, do it now. Do not wait. All areas in the national forests are first-come, first-served. Forest roads will be packed. Expect all areas to be swarming with people. Show up early to stake out your spot. Consider staying farther away and driving early on August 21.

Please respect private land too. Oregon folks don't take kindly to people overrunning their property without permission. In a big state with only 3.97 million residents, people are very protective, but they're friendly, too. You never know what you might be able to arrange with a smile and a bit of money.

This all said, there are plenty of camping opportunities throughout Oregon. You don't have to sleep exactly on the eclipse path. If you're ready to rough it, there are national forest camping options.

Government agencies have been meeting since 2015 to talk about how to manage the influx of people. Please note that every possible government agency will be working full time to enforce the various rules and regulations.

National Monuments

Chances are you will not find a camping site at any state park, national park, or national monument in Oregon. To watch the eclipse from any location, you do not have to sleep in it. You just need to drive there in the morning.

Park law enforcement will be present on the eclipse weekend. Hundreds of thousands of people are expected in the region. Parking will overflow. It will make parking lots and lines on Black Friday at the mall look uncrowded. For an event of this magnitude, you'll need to find your parking space early.

The first sentence of the national parks mission statement is:

"The National Park Service preserves unimpaired the natural and cultural resources and values of the national park system for the enjoyment, education, and inspiration of this and future generations."

Roadside camping (sleeping in your car) is not allowed in national monuments or parks. Park facilities are only designed to handle so many people per day. Water, trash collection, and toilets can only withstand so much. If you notice trash on the ground, take a moment to throw it away. Protect your national park and help out. Rangers are diligent and hardworking but they can only do so much to manage the expected crowds.

National Forests

There are plenty of national forests in Oregon. They all have camping opportunities. The forest service manages undeveloped and primitive campsites. Be sure to check for any fire restrictions. Check with individual agencies for last-minute information and regulations. The forest service requires proper food storage. Plan to purchase food and water before choosing your campsite. Below is a partial list of national forests along or near the total eclipse path:

Malheur NF:	www.fs.usda.gov/malheur
Ochoco NF:	www.fs.usda.gov/ochoco
Umatilla NF:	www.fs.usda.gov/umatilla
Willamette NF:	www.fs.usda.gov/willamette

Forest service roads abound in Oregon. Maps for forests are available at local visitor centers and bookstores. This book's website has digital copies of some forest maps.

Printed national forest maps are large and detailed. They have illustrated road paths, connections, and other vital travel information not available on digital device maps. Viewing digital maps on your smartphone or pad is difficult. If you plan to camp in the forest, a real paper map is a wise investment.

Camping in federal wilderness areas is also allowed. Those areas afford the ultimate backcountry experience. However, be aware that no vehicle travel is allowed in the specially designated areas. This ban includes: vehicles, bikes, hang gliders, and drones. You can travel only on foot or with pack animals.

Warm Springs Indian Reservation

The total eclipse will completely cover the Warm Springs Reservation near Warm Springs. NASA is working with the Native American middle school and high school teams for a special science project. They are planning to launch high-altitude balloon payloads to view the moon's shadow from the edge of space. Read more about this eclipse project here:

wsuas.com/2016/09/nasa-partner-solar-eclipse-2017

www.waspacegrant.org/2017eclipse

Visit the reservation's webpage for more information about access, travel, and eclipse viewing: warmsprings-nsn.gov.

Bureau of Indian Affairs: www.bia.gov

Sleep in Your Car

Countless RVs, campers, trucks, cars, and motorcycles will flood Oregon. Sleeping in your car with friends is tolerable. Doing so with unadventurous spouses or children is another matter.

Do not be caught along the path of the total eclipse without some sort of plan, especially in coastal Oregon. The whole area brims with people on a normal summer day. August 21 will be anything but normal.

Useful Local Webcams

Local webcams are handy to make last-minute travel decisions. The webcams are sensitive enough to show headlights at night. Use them to determine if there are issues before traveling out. Eclipse traffic will add to the morning commuter traffic.

The smartphone application Wunderground is useful to check on webcams in one place. All the webcams are listed near the bottom of the app window.

Weather

It's all about the weather during the eclipse. Nothing else will matter if the sky is cloudy. You can be nearly anywhere in Oregon and catch a view of the sky when traffic comes to a standstill. But if there's a cloud-cover forecast, seriously reconsider your viewing location.

Travel early wherever you plan to go. Attempting to change locations an hour before the eclipse due to weather will likely cause you to miss the event. Oregon mountain roads are narrow and slow. The mountain passes are steep, and vehicles back up routinely.

MODERN FORECASTS

Use a smartphone application to check the up-to-date weather. Wunderground is a good application and has relatively reliable forecasts for the region. The hourly forecast for the same day has been rather accurate for the last two years. The below discussion refers to features found in the Wunderground app. However, any application with detailed weather views will improve your eclipse forecasting skills.

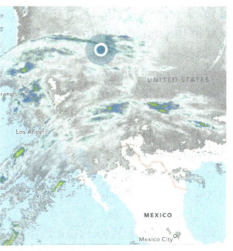
Infrared cloud map showing the worst case eclipse cloud cover.
Courtesy of National Weather Service.

CLOUD COVER FORECAST

The most useful forecast view is the visible and infrared cloud-coverage map. Avoid downloading this app the night before and trying to learn how to read it. Practice reading them at home. It's imperative to understand how to interpret the maps early.

All cloud cover, night or day, will appear on an infrared map. Warm, low-altitude clouds are shown in white and gray. High-altitude cold clouds are displayed in shades of green, yellow, red, and purple. Anything other than a clear map spells eclipse-viewing problems.

To improve your weather guess, use the animated viewer of the cloud cover. It will give you a sense of cloud motion. You can discern whether clouds or rain are moving toward, away from, or circulating around your location.

Normal Oregon Weather Pattern

Due to the direction of the jet stream, most weather travels across the Pacific Ocean and into Oregon. It then flows on to Idaho and Montana. On occasion, weather can approach from any direction. Due to the nature of the coastal mountains, weather in Oregon can be unpredictable.

The common weather pattern in August is clear in the morning and cloudy in the afternoon in the western half of the state. The eastern half of the state tends to be dryer.

Cities east of any Oregon mountain ranges tend to have clear skies during August. Do not rely on this action though. If anything other than clear skies are predicted, plan to drive to other parts of Oregon, Idaho, Wyoming, or even Nebraska.

Consider that slow-moving clouds can obscure the sun for far longer than the two-minute duration of the totality. The time of totality is so short that you do not want to risk it. Missing it due to a single cloud will be a major disappointment.

Local Eclipse Weather Forecasts

Local town and city newspapers, radio, and television stations around Oregon will have a weekend edition with articles discussing the eclipse weather. However, conditions change unpredictably in the coastal mountain range. A three-day forecast may be completely incorrect.

Forest Fires

For the past several years, forest fires have been common in the western United States. The summer of 2017 is likely to be no different. There were major fires in Oregon during the summer of 2015. Chances are there will be fires again in the region in the summer of 2017. For fire updates check https://inciweb.nwcg.gov.

For the best eclipse viewing experience, you need to have as clear a sky as possible. Fog, clouds, or smoke will obscure the subtleties of the sun's corona. If you think the view of the sky is going to be blocked, don't wait until the last minute to move to a clearer location. If you wait too long to decide to move to a better viewing area, it may be impossible due to traffic.

Road Closures Due to Fires

Highways connecting various Oregon towns were closed for several days during the summer of 2015. The Canyon Creek Fire turned into a monster blaze. This fire was created after the Mason Spring and Berry Creek fires met.

In 2016, Oregon enjoyed a relatively quiet fire season. That year, 5,554 acres burned that year, costing over $17 million to fight. With the heavy snowpack from multiple powerful winter storms, it's a guess what will happen in August 2017. If forest areas continue to remain wet, the whole region may dodge the fires of years previous.

Although fire is an important part of forest ecology, it worries eclipse chasers. Other than clouds, smoke from fires will block the view of the sun and moon on the morning of the eclipse. Should there be fires where you are or may be headed, reconsider your location as early as possible. The most accurate website for fires is:

inciweb.nwcg.gov

Check the Oregon road report for updated closure information:
www.tripcheck.com

It's imperative to plan for fires and their effects. Watch the weather reports. If strong winds and lightning storms are forecast, prepare to

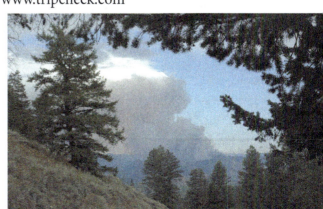

change your viewing location. If conditions are poor, you and thousands of other vehicles will be trapped in slow-moving traffic.

If you believe it's necessary to leave a town to watch the eclipse, do so the night before or extremely early in the morning. RVs are common, and trains of them crawl over mountain passes.

Oregon Information

Cellular Phones

Cellular "cell" phone service in remote Oregon locations is spotty at best. Most of the time there is good coverage along the main highways and interstates. However, even along major thoroughfares, there can be little or no coverage.

It's possible to find zones where text messages will send when phone calls are impossible. If you cannot make a phone call, the chance of having data coverage for web surfing or e-mail is low.

Please look up any information or communicate what you need before departing from the main roads around Oregon. Bureau of Land Management (BLM) areas sometimes have coverage. Planned to be self-contained. Treat like your cell phone like it won't connect.

You may find yourself out of cell service. With a large number of cell users in a concentrated area, coverage and data speed may collapse as well. Search on the phrase "cell phone coverage breathing".

Wilderness and Forest Safety

Nearly all regions of Oregon are full of wild animals. Although they are beautiful, the animals are dangerous. They can easily injure or kill people, as they are far more powerful than humans. Do not try to feed any wild animals, including squirrels, foxes, and chipmunks, as they can carry diseases. These suggestions apply to all public lands.

Bear

The forests of Oregon are home to black bears. Safety is imperative around these powerful animals. Although they often appear docile, they can become aggressive if threatened. In the unlikely event of an attack, fight back against the bear. Use whatever you have at your disposal to defend yourself. Report all negative or aggressive bears to the local authorities.

If a bear hears you, it will usually vacate the area. Bear charges are often caused by unexpected and surprise encounters. Noise is the best defense to avoid surprising bears. Regularly clap, make noise, and talk loudly. The *Oregon Department of Fish and Wildlife Bear Safety Brochure* has more specific information on safety and food management in bear country.

It is recommended to stay one hundred yards (300 feet) away from all bears. They are exciting to see but need their space. Refer to current forest or park regulations for more safety information.

Rattlesnakes

There are two primary species of rattlesnake in Oregon: Great Basin Rattlesnake and the Western Rattlesnake (Northern Pacific Rattlesnake). Although these reptiles are not generally aggressive, they can strike when provoked or threatened. Of the approximately 8,000 people annually bitten by venomous snakes in the United States, ten to fifteen people die according to the U.S. Food and Drug Administration.

The best way to avoid rattlesnake encounters is to be mindful of your environment. Do not place your hands or feet in locations where you cannot clearly see the surroundings. Avoid heavy brush or tall weeds where snakes hide during the day. Do not wander around at night without a light source. Step on a log or rock rather than over it, as a hidden snake might be on the other side. Rattlesnakes may not make any noise before striking.

Avoid handling all rattlesnakes. Although online media has popularized dangerous snake handling, this dramatically increases the chances of being injured. If you are bitten, stay calm and call 911 or emergency dispatch as soon as possible. Transport the victim to the nearest medical facility immediately. Rapid professional treatment is the best way to manage rattlesnake bites. Refer to U.S. Forest Service and professional medical texts for more information on managing rattlesnakes injuries.

Ticks

Ticks exist all across the United States, but not all species transmit disease. Ticks cannot fly or jump, but they climb grasses in shrubs in order to attach to people or animals that pass by. Ticks feed on the blood of their host. In doing so, they can transmit potentially life-threatening diseases such as Lyme disease.

According to the National Pesticide Information Center, ticks must be attached and feed for several hours before an infection can be passed. Do not wait that long to manage tick exposure. As soon as you travel through outdoor locations, especially in grasses or shrubs, check yourself for ticks. They can be incredibly small and difficult to detect. Some are smaller than the head of a pin.

You may wish to consider wearing tick-specific insect repellent. Check with your doctor or medical professional about any potential adverse side effects of chemical repellents.

Should you discover an attached tick, follow modern and current tick removal methods. There are many incorrect tick removal methods found online. Refer to your medical professional for proper removal methods, as improper removal can result in an infection.

Scorpions

Scorpions are generally found in the dry desert areas of Oregon. However, they may be anywhere in the state. Avoid all contact with these arthropods, as their sting is painful and dangerous. For more information, visit the CDC webpage: www.cdc.gov/niosh/topics/insects.

Mountain Lion

Although exciting and beautiful, an encounter with a mountain lion can be a frightening experience. Should you encounter a mountain lion, do not run. It may stimulate a predatory response from the cat. Keep calm, back away slowly, and maintain eye contact. Do all you can to appear larger. Stand upright, raise your arms, or hoist your jacket. Never bend over or crouch down.

If attacked, fight back, as your life depends on it. Use anything you have at your disposal. Pick up small children if you spot a mountain lion. They perceive children as prey. Pick them up. For more moun-

tain lion information, visit the Oregon Department of Fish and Wildlife webpage:
www.dfw.state.or.us/wildlife/living_with/cougars.asp

Eclipse Day Safety

1. Hydrate

Summers can be extremely warm. The excitement of the event can distract you from managing hydration. Drink plenty of water. Consume more than you would at home, especially if you live in a humid environment.

2. Eye Safety time

Use certified eclipse safety glasses at all times when viewing the partial eclipse. Only remove the glasses when the totality happens. Give your eyes time to rest. Even though the glasses are safe, your eyes can dry out and become irritated. Bring FDA-approved eye drops to keep your eyes moist.

3. Sun exposure

The sun is much more intense in August than at other times of the year. Wear sunglasses, and liberally apply sunscreen to avoid sunburns.

4. Eat well

Keep your energy up. Appetite loss is common when traveling. Maintain your normal eating schedule.

5. Prepare for temperature changes

Temperatures will drop rapidly once the sun sets, especially in the mountains or desert. Bring appropriate clothing.

6. Talk with your doctor

If the altitude, travel, or stress bothers you talk with your doctor before traveling. Seek professional medical attention if you develop serious symptoms.

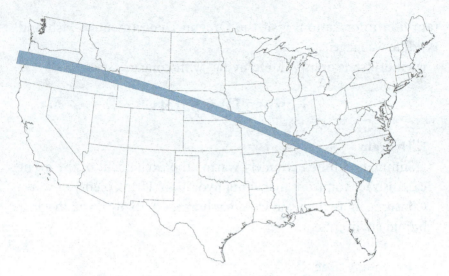

Total eclipse path across the United States (approximate).

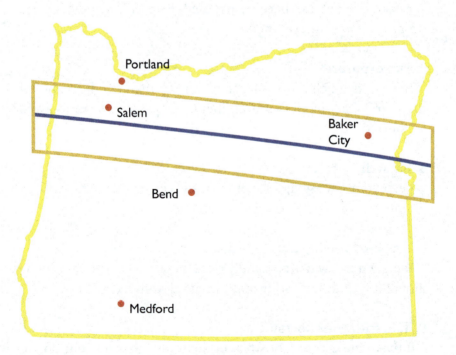

Total eclipse path across Oregon (approximate).

All About Eclipses

How an Eclipse Happens

An eclipse occurs when one celestial body falls in line with another, thus obscuring the sun from view. This occurs much more often than you'd think, considering how many bodies there are in the solar system. For instance, there are over 150 moons in the solar system. On Earth, we have two primary celestial bodies: the sun and the moon. The entire solar system is constantly in motion, with planets orbiting the sun and moons orbiting the planets. These celestial bodies often come into alignment. When these alignments cause the sun to be blocked, it is called an eclipse.

For an eclipse to occur, the sun, Earth, and moon must be in alignment. There are two types of eclipses: solar and lunar. A solar eclipse occurs when the moon obscures the sun. A lunar eclipse occurs when the moon passes through Earth's shadow. Solar eclipses are much more common, as we experience an average of 240 solar eclipses a century compared to an average of 150 lunar eclipses. Despite this, we are more likely to see a lunar eclipse than a solar eclipse. This is due to the visibility of each.

For a solar eclipse to be visible, you have to be in the moon's shadow. The problem with viewing a total eclipse is that the moon casts a small shadow over the world at any given time. You have to be in

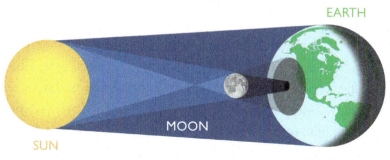

* ILLUSTRATION NOT TO SCALE

a precise location to view a total eclipse. The issue that arises is that most of these locations are inaccessible to most people. Though many would like to see a total solar eclipse, most aren't about to set sail for the middle of the Pacific Ocean. In fact, a solar eclipse is visible in the same place on the world on average every 375 years. This means that if you miss a solar eclipse above your hometown, you're not going to see another one unless you travel or move.

It's much easier to catch a glimpse of a lunar eclipse, even though they occur at a much lower frequency than their solar counterparts. A lunar eclipse darkens the moon for a few hours. This is different than a new moon when it faces away from the sun. During these eclipses, the moon fades and becomes nearly invisible.

Another result of a lunar eclipse is a blood moon. Earth's atmosphere bends a small amount of sunlight onto the moon turning it orange-red. The blood moon is caused by the dawn or dusk light being refracted onto the moon during an eclipse.

Lunar eclipses are much easier to see. Even when the moon is in the shadow of Earth, it's still visible throughout the world because of how much smaller it is than Earth.

Total vs. Partial Eclipse

What is the difference between a partial and total eclipse? A total eclipse of either the sun or the moon will occur only when the sun, Earth, and the moon are aligned in a perfectly straight line. This ensures that either the sun or the moon is partially or completely obscured.

In contrast, a partial eclipse occurs when the alignment of the three celestial bodies is not in a perfectly straight line. These types of eclipses usually result in only a part of either the sun or the moon being obscured. This is often what led to ancient civilizations believing that some form of magical beast or deity was eating the sun or the moon. It appears as though something has taken a bite out of either the sun or the moon during a partial eclipse.

Total eclipses, rarer than partial eclipses, still occur quite often. It's more difficult for people to be in a position to experience such an event firsthand. Total solar eclipses can only be viewed from a small portion of the world that falls into the darkest part of the moon's shadow. Often this happens in the middle of the ocean.

THE MOON'S SHADOW

The moon's shadow is divided into two parts: the umbra and the penumbra. The former is much smaller than the latter, as the umbra is the innermost and darkest part of the shadow. The umbra is thus the central point of the moon's shadow, meaning that it is extremely small in comparison to the entire shadow. For a total solar eclipse to be visible, you need to be directly beneath the umbra of the moon's shadow. This is because that is the only point at which the moon completely blocks the view of the sun.

In contrast, the penumbra is the region of the moon's shadow in which only a portion of the light cast by the sun is obscured. When

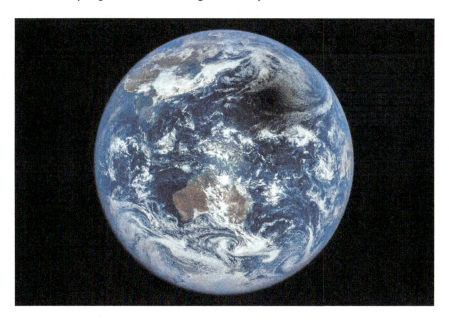

Total eclipse shadow 2016 as seen from 1 million miles on the Deep Space Climate Observatory satellite. Courtesy of NASA.

standing in the penumbra, you are viewing the eclipse at an angle. In the penumbra, the moon does not completely block the sun from view. This means that while the event is a total solar eclipse, you'll only see a partial eclipse. The umbra for the August 21st eclipse is approximately sixty miles wide. The penumbra will cover much of the United States.

To provide some context, the last total solar eclipse we experienced occurred on March 9, 2016, and was visible as a partial eclipse across most of the Pacific Ocean, parts of Asia, and Australia. However, the only place in the world to view this total solar eclipse was in a few parts of Indonesia.

Due to the varied locations and the brief periods for which they're visible, it's difficult to see each and every eclipse that occurs. Many people don't even realize that they have occurred. Consider that the umbra of the moon represents such a small fraction of the entire shadow and the majority of our planet is comprised of water. Thus, the rarity of being able to view a total solar eclipse increases significantly because it's likely that the umbra will fall over some part of the ocean rather than a populated landmass.

Eclipses Throughout History

Ancient peoples believed eclipses were from the wrath of angry gods, portents of doom and misfortune, or wars between celestial beings. Eclipses have played many roles in cultures, creating myths since the dawn of time. Both solar and lunar eclipses affected societies worldwide. Inspiring fear, curiosity, and the creation of legends, eclipses have cast a long shadow in the collective unconscious of humanity throughout history.

Early Myth & Astronomy

Documented observations of solar eclipses have been found as far back in history as ancient Egyptian and Chinese records. Timekeeping was important to ancient Chinese cultures. Astronomical observations were an integral factor in the Chinese calendar. The first observation of a solar eclipse is found in Chinese records from over 4,000 years ago. Evidence suggests that ancient Egyptian observations may predate those archaic writings.

Many ancient societies, including Roman, Greek and Chinese civilizations, were able to infer and foresee solar eclipses from astronomical data. The sudden and unpredictable nature of solar eclipses had a stressful and intimidating effect on many societies that lacked the scientific insight to accurately predict astronomical events. Relying on the sun for their agricultural livelihood, those societies interpreted solar eclipses as world-threatening disasters.

In ancient Vietnam, solar eclipses were explained as a giant frog eating the sun. The peasantry of ancient Greece believed that an eclipse was the sign of a furious godhead, presenting an omen of wrathful retribution in the form of natural disasters. Other cultures were less speculative in their investigations. The Chinese Song Dynasty scientist Shen Kuo proved the spherical nature of the Earth and heavenly bodies through scientific insight gained by the study of eclipses.

The Eclipse in Native American Mythology

Eclipses have played a significant role in the history of the United States. Before Europeans settled in the Americas, solar eclipses were important astronomical events to Native American cultures. In most native cultures, an eclipse was a particularly bad omen. Both the sun and the moon were regarded as sacred. Viewing an eclipse, or even being outside for the duration of the event, was considered highly taboo by the Navajo culture. During an eclipse, men and women would simply avert their eyes from the sky, acting as though it was not happening.

The Choctaw people had a unique story to explain solar eclipses. Considering the event as the mischievous actions of a black squirrel and its attempt to eat the sun, the Choctaw people would do their best to scare away the cosmic squirrel by making as much noise as possible until the end of the event, at which point cognitive bias would cause them to believe they'd once again averted disaster on an interplanetary scale.

Contemporary American Solar Phenomena

The investigation of solar phenomena in twentieth-century American history had a similarly profound effect on the people of the United States. A total solar eclipse occurring on the sixteenth of June, 1806, engulfed the entire country. It started near modern-day Arizona. It passed across the Midwest, over Ohio, Pennsylvania, New York, Massachusetts, and Connecticut. The 1806 total eclipse was notable for being one of the first publicly advertised solar events. The public was informed beforehand of the astronomical curiosity through a pamphlet written by Andrew Newell entitled *Darkness at Noon, or the Great Solar Eclipse*.

This pamphlet described local circumstances and went into great detail explaining the true nature of the phenomenon, dispelling myth and superstition, and even giving questionable advice on the best methods of viewing the sun during the event. Replete with a short historical record of eclipses through the ages, the *Darkness at Noon* pamphlet is one of the first examples of an attempt to capitalize on the mysterious nature of solar eclipses.

Another notable American solar eclipse occurred on June 8, 1918. Passing over the United States from Washington to Florida, the eclipse was accurately predicted by the U.S. Naval Observatory and heavily documented in the newspapers of the day. Howard Russell Butler, painter and founder of the American Fine Arts Society, painted the eclipse from the U.S. Naval Observatory, immortalizing the event in *The Oregon Eclipse*.

Four more total solar eclipses occurred over the United States in the years 1923, 1925, 1932, and 1954, with another occurring in 1959. The October 2, 1959, solar eclipse began over Boston, Massachusetts. It was a sunrise event that was unviewable from the ground level. Eminent astronomer Jay Pasachoff attributed this event to sparking his interest in the study of astronomy. Studying under Professor Donald Menzel of Williams College, Pasachoff was able to view the event from an airline hired by his professor.

To this day, many myths surround the eclipse. In India, some local customs require fasting. In eastern Africa, eclipses are seen as a danger to pregnant women and young children. Despite the mystery and legend associated with unique and rare astronomical events, eclipses continue to be awe-inspiring. Even in the modern day, eclipses draw out reverential respect for the inexorable passing of celestial bodies. They are a reminder of the intimate relationship between the denizens of Earth and the universe at large.

Present Day Eclipses

The year 2016 brought the world just two solar eclipses. A total solar eclipse occurred on the 9th of March. An annular solar eclipse, in which the sun appears as a "ring of fire" occurred on the 23rd of March. If you're interested in seeing this rare and exciting solar

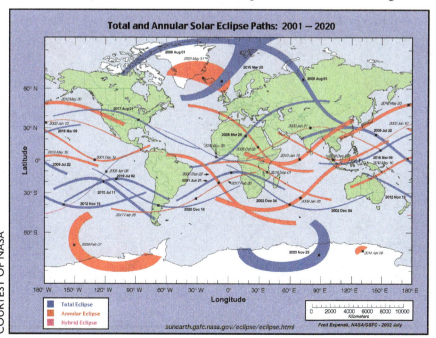

phenomenon yourself, you must to travel to either South America or Western Africa on the 26th of February, 2017.

The next total solar eclipse viewable from the United States, or anywhere else in the world, will occur on the 21st of August, 2017. It will be visible in Oregon, Idaho, Wyoming, Nebraska, Missouri, Kentucky, Tennessee, and South Carolina. The event will be the only total solar eclipse for Americans this decade.

Future American Eclipses

The next total eclipse to cross the continental United States is on April 8, 2024. It will travel from Texas to Maine. After that, the next American total eclipses will be in 2044 and 2045. The next partial eclipse to touch Oregon will be in 2023. You will have to wait until 2169 for the next total eclipse to pass over Oregon.

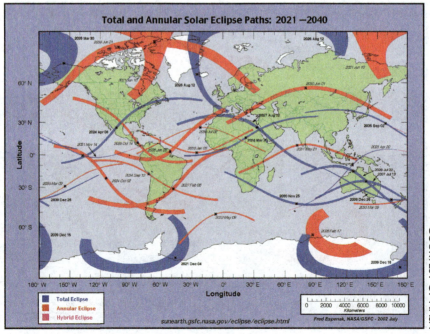

Viewing and Photographing the Eclipse

AT-HOME PINHOLE METHOD

Use the pinhole method to view the eclipse safely. It costs little but is the safest technique there is. Take a stiff piece of single-layer cardboard and punch a clean pinhole. Let the sun shine through the pinhole onto another piece of cardboard. That's it!

Never look at the sun through the pinhole. Your back should be toward the sun to protect your eyes. To brighten the image, simply move the back piece of cardboard closer to the pinhole. To see it larger, move the back cardboard farther away. Do not make the pinhole larger. It will only distort the crescent sun.

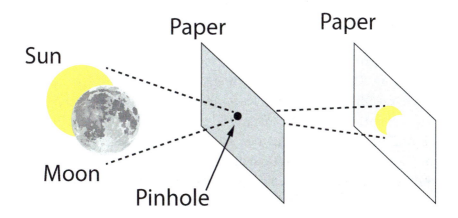

WELDING GOGGLES

Welding goggles that have a rating of fourteen or higher are another useful eclipse viewing tool. The goggles can be used to view the solar eclipse directly. Do not use the goggles to look through binoculars or telescopes, as the goggles could potentially shatter due to intense direct heat. Avoid long periods of gazing with the goggles. Look away every so often. Give your eyes a break.

SOLAR FILTERS FOR TELESCOPES

The ONLY safe way to view solar eclipses using telescopes or binoculars is to use solar filters. The filters are coated with metal to diminish the full intensity of the sun. Although the filters can be expensive, it is better to purchase a quality filter rather than an inexpensive one that could shatter or melt from the heat.

The filters attach to the front of the telescope for easy viewing. Remember to give your telescope cooling breaks. Rapid heating can damage your equipment with or without filters attached.

Watch Out for Unsafe Filters

There are several myths surrounding solar filters for eclipse viewing. In order for filters to be safe, they must be specially designed for looking at a solar eclipse. The following are all unsafe for eclipse viewing and can lead to retinal damage: developed colored or chromogenic film, black-and-white negatives such as X-rays, CDs with aluminum, smoked glass, floppy disk covers, black-and-white film with no silver, sunglasses, or polarizing films.

Some online articles state that using developed black-and-white film is safe. Those articles fail to mention the film must have a layer of real metallic silver to protect your eyes. Using developed film is discouraged. You cannot ensure the quality of the film. Feeling no discomfort while looking at the partial solar eclipse does NOT mean your eyes are protected. Retinal damage can occur with zero pain due to the retinas having no pain receptors. Please be careful. Only use protective glasses certified for viewing the eclipse.

Viewing with Binoculars

When viewing the eclipse with binoculars, it is important to use solar filters on both lenses until totality. Only then is it safe to remove the filter. As the sun becomes visible after totality, replace the filters for safe viewing. Protect your pupils. Remember to give your binoculars a cool-down break between viewings. They can overheat rapidly from being pointed directly at the sun even with filters attached.

Planning Ahead

There are many things to keep in mind when viewing a total eclipse. It is important to plan ahead to get the most out of this extraordinary experience.

UNDERSTANDING SUN POSITION

All compass bearings in this book are true north. All compasses point to Earth's magnetic north. The difference between these two measurements is called magnetic declination. The magnetic declination for Oregon in August 2017 is:

$$15°\ 15'\ E\ \pm\ 0°\ 21'\ \text{(for Salem)}$$

Subtract the declination from the azimuth bearing as given in the text, and set your compass to that direction.

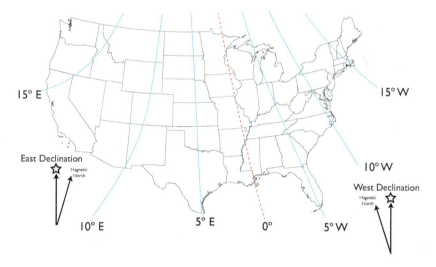

If you purchase a compass with a built-in declination adjustment, you can change the setting once and eliminate the calculations. The Suunto M-3G compass has this correction. A compass with a sighting mirror or wire will help you make a more accurate azimuth sighting.

The Suunto M-3G also has an inclinometer. This allows you to measure the elevation of any object above the horizon. Use this to figure out how high the sun will be above your position.

You can also use a smartphone inclinometer and compass for this purpose. Make sure to calibrate your smartphone's compass before every use, otherwise it might indicate the wrong bearing. Set the smartphone compass for true north to match the book. Understand the compass prior to August 21. There will be little time to guess or search on Google. Smartphone and GPS compasses are "sticky." Their compasses don't swing as freely as a magnetic compass does.

The author has used his magnetic compass for azimuth measurements and a smartphone to measure elevation. Combining these two tools will allow you to make the best sightings possible.

Outdoor sporting goods stores in most Oregon towns and cities carry compasses. We recommend purchasing a good compass in your hometown. Take the time to learn how to use it before the day of the eclipse. You do not want to struggle with orienteering basics under pressure.

Sun Azimuth

Azimuth is the compass angle along the horizon, with 0° corresponding to north, and increasing in a clockwise direction. 90° is east, 180° is south, and 270° is west.

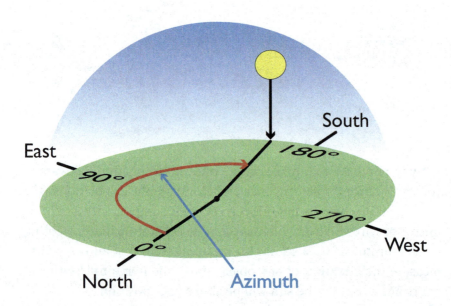

Sun Elevation

Altitude is the sun's angle up from the horizon. A 0° altitude means exactly on the horizon and 90° means "straight up."

Using the sun azimuth and elevation data, you can predict the position of the sun at any given time. Positions given in this book coincide with the time of eclipse totality unless otherwise noted.

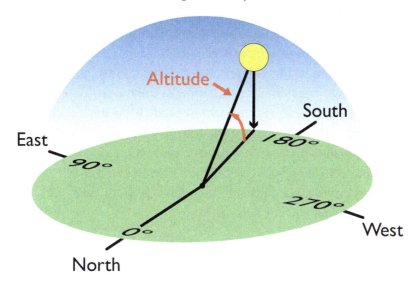

ECLIPSE DATA FOR SELECT OREGON LOCATIONS

LOCATION	TOTALITY START (PDT)	ALTITUDE	AZIMUTH
ALBANY	10:17:05AM	39°	116°
BAKER CITY	10:24:31AM	41°	124°
LINCOLN CITY	10:16:07AM	38°	115°
MADRAS	10:19:36AM	41°	119°
NEWPORT	10:15:55AM	39°	115°
SALEM	10:17:21AM	39°	116°

ECLIPSE PHOTOGRAPHY

Photographing an eclipse is an exciting challenge, as the moon's shadow moves near 2,000MPH. There is an element of danger and the pressure of time. Looking at the unfiltered sun through a camera can permanently damage your vision and your camera. If you are unsure, just enjoy the eclipse with specially designed glasses. Keep a solar filter on your lens during the eclipse and remove for the duration of totality!

Partial Vs. Total Solar Eclipse

To successfully and safely photograph a partial and total eclipse, it is important to understand the difference between the two. A solar eclipse occurs when the moon is positioned between the sun and Earth. The region where the shadow of the moon falls upon Earth's surface is where a solar eclipse is visible.

The moon's shadow has two parts—the penumbral shadow and the umbral shadow. The penumbral shadow is the moon's outer shadow where partial solar eclipses can be observed. Total solar eclipses can only be seen within the umbral shadow, the moon's inner shadow.

You cannot say you've seen a total eclipse when all you saw was a partial solar eclipse. It is like saying you've watched a concert, but in reality, you only listened outside the arena. In both cases, you have missed the drama and the action.

Photographing A Partial And Total Solar Eclipse

Aside from the region where the outer shadow of the moon is cast, a partial solar eclipse is also visible before a total solar eclipse within

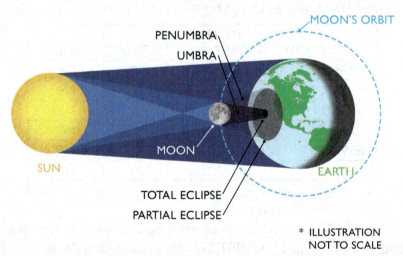

the inner shadow region. In both cases, it is imperative to use a solar filter on the lens for both photography and safety reasons. This is the only difference between taking a partial eclipse and a total eclipse photograph of the sun.

To photograph a total solar eclipse, you must be within the Path of Totality, the surface of the Earth within the moon's umbral shadow.

The Challenge

A total solar eclipse only lasts for a couple of minutes. It is brief, but the scenario it brings is unforgettable. Seeing the radiant sun slowly being covered by darkness gives the spectator a high level of anticipation and indescribable excitement. Once the moon completely covers the sun's radiance, the corona is finally visible. In the darkness, the sun's corona shines, capturing the crowd's full attention. Watching this phenomenon is a breathtaking experience.

Amidst all the noise, cheering, and excitement, you have less than 150 seconds to take a perfect photograph. The key to this is planning. You need to plan, practice, and perfect what you will do when the big moment arrives because there is no replay. The pressure is enormous. You only have two minutes to capture the totality and the sun's corona using different exposures.

Plan, Practice, Perfect

It is important to practice photographing before the actual phenomenon arrives. Test your chosen imaging setup for flaws. Rehearse over and over until your body remembers what you will do from the moment you arrive at your chosen spot to the moment you pack up and leave the area.

You will discover potential problems regarding vibrations and focus that you can address immediately. This minimizes the variables that might affect your photographs at the most critical moment.

It's common for experienced eclipse chasers to lose track of what they plan to do. Write down what you expect to do. Practice it time and again. Play annoying, distracting music while you practice. Try photographing in the worst weather possible. Do anything you can to practice under pressure. Eclipse day is not the time to practice.

Once the sun is completely covered, don't just take photographs. Capture the experience and the image of the total solar eclipse in your mind as well. Set up cameras around you to record not just the total solar eclipse but also the excitement and reaction of the crowd.

Eclipse Photography Gear

What do you need to photograph the total eclipse? There are only a few pieces of equipment that you'll need. Preparing to photograph an eclipse successfully takes time. Not only do you have to be skilled and have the right gear, you have to be in the correct place.

Basic Eclipse Photography Equipment
- Solar viewing glasses
- Lens solar filter
- Minimum 300mm lens
- Stable tripod that can be tilted to 60° vertical
- High-resolution DSLR
- Spare batteries for everything
- Secondary camera to photograph people, the horizon, etc.
- Remote cable or wireless release

Additional Items
- Video camera
- Video camera tripod
- Quality pair of binoculars
- Solar filters for each binocular lens
- Photo editing software

Equipment to Prepare Before the Big Day

A. Solar viewing glasses
You need a pair of solar viewing glasses as the eclipse approaches.

B. Solar Filter
Partial and total eclipse photography is different from normal photography. Even if only 1% of the sun's surface is visible, it is still approximately 10,000 times brighter than the moon. Before totality, use a solar filter on your lens. Do not look at the sun with your eyes. It can cause irreparable damage to your retinas.

DO NOT leave your camera pointed at the sun without a solar filter attached. The sun will melt the inside of your camera. Think of a magnifying glass used to torch ants and multiply that by one hundred.

C. Lens

To capture the corona's majesty, you need to use a telescope or a telephoto lens. The best focal length, which will give you a large image of the sun's disk, is 400mm and above. You don't want to waste all your efforts by bringing home a small dot where the black disk and majestic corona are supposed to be.

D. Tripod

Bring a stable enough tripod to support your camera properly to avoid unsteady shots and repeated adjustments. Either will ruin your photos. It also needs to be portable in case you need to change locations for a better shot.

E. Camera

You need to remember to set your camera to its highest resolution to capture all the details. Set your camera to:

- 14-bit RAW is ideal, otherwise
- JPG, Fine compression, Maximum resolution

Bracket your exposures. Shoot at various shutter speeds to capture different brightnesses in the corona. Note that stopping your lens all the way down may not result in the sharpest images.

Choose the lowest possible ISO for the best quality while maintaining a high shutter speed to prevent blurred shots. Set your camera to manual. Do not use AUTO ISO. Your camera will be fooled. The night before, test the focus position of your lens using a bright star or the moon.

Constantly double-check your focus. Be paranoid about this. You can deal with a grainy picture. No amount of Photoshop will fix a blurry, out-of-focus picture.

F. Batteries

Remember to bring fresh batteries! Make sure that you have enough power to capture the most important moments. Swap in fresh batteries thirty minutes before totality.

G. Remote release

Use a wired or wireless remote release to fire the camera's shutter. This will reduce the amount of camera vibration.

H. Video Camera

Run a video camera of yourself. Capture all the things you say and do during the totality. You'll be amazed at your reaction.

I. Photo editing software

You will need quality photo editing software to process your eclipse images. Adobe Lightroom and Photoshop are excellent programs to extract the most out of your images. Become well versed in how to use them at least a month before the eclipse.

J. Smartphone applications

The following smartphone applications will aid in your photography planning: Wunderground, Skyview, Photographer's Ephemeris, Sunrise and Sunset Calculator, SunCalc, and Sun Surveyor among others.

Camera Phones

Smartphone cameras are useful for many things but not eclipse photography. An iPhone 6 camera has a 63° horizontal field of view and is 3264 pixels across. If you attempt to photograph the eclipse, the sun will be a measly 30-40 pixels wide depending on the phone. Digital pinch zoom won't help here. If you want *National Geographic* images, you'll need a serious camera and lens, far beyond any smartphone.

Consider instead using a smartphone to run a time-lapse of the entire event. The sun will be minuscule when shot on a smartphone. Think of something else exciting and interesting do to with it. Purchase a Gorilla Pod, inexpensive tripod, or selfie stick and mount the smartphone somewhere unique.

Also, partial and total eclipse light is strange and ethereal. Consider using that light to take unique pictures of things and people. It's rare and you may have something no one else does.

Focal Length & the Size of Sun

The size of the sun in a photo depends on the lens focal length. A 300mm lens is the recommended minimum on a full-frame (FF) DSLR. Lenses up to this size are relatively inexpensive. For more magnification, use an APS-C (crop) size sensor. Cameras with these sensors provide an advantage by capturing a larger sun.

For the same focal length, an APS-C sensor will provide a greater apparent magnification of any object. As a consequence, a shorter, less expensive lens can be used to capture the same size sun.

The below figure shows the size of the sun on a camera sensor at various focal lengths. As can be seen with the 200mm lens, the sun is quite small. On a full-frame camera at 200mm, the sun will be 371 pixels wide on a Nikon D810, a 36-megapixel body. A lower resolution FF camera will result in an even smaller sun.

Printing a 24-inch image shot on a Nikon D810 with a 200mm lens at a standard 300 pixels per inch results in a small sun. On this size paper, the sun will be a miserly 1.25 inches wide!

Photographing the eclipse with a lens shorter than 300mm will leave you with little to work with. Using a 400mm lens and printing a 24-inch print will result in a 2.5-inch-wide sun. For as massive as the sun is, it is a challenge to take a photograph with the sun of any meaningful size.

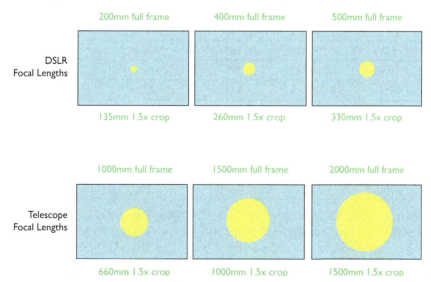

FOCAL LENGTH	FOV FULL FRAME	FF VERT. ANGLE	% OF FF	SUN PIXEL SIZE
14	104° X 81°	81°	0.7%	32.1
20	84° X 62°	62°	0.9%	41.9
28	65° X 46°	46°	1.2%	56.5
35	54° X 38°	38°	1.4%	68.5
50	40° X 27°	27°	2.0%	96.4
105	19° X 13°	13°	4.1%	200.2
200	10° X 7°	7°	7.6%	371.9
400	5° X 3.4°	3.4°	15.6%	765.6
500	4° X 2.7°	2.7°	19.6%	964.2
1000	2° X 1.3°	1.3°	40.8%	2002.5
1500	1.4° X 0.9°	0.9°	58.9%	2892.6
2000	1° X 0.68°	0.68°	77.9%	3828.4

Chart 1: Full-frame camera field of view. The 3rd column is the vertical field of view in degrees. Column 4 is the percentage of the total sensor height that the sun covers. Column 5 is how many pixels wide the sun will be on a 36MP Nikon D810. (Values are estimates)

FOCAL LENGTH	FOV CROP	CROP VERT DEG	% OF CROP	SUN PIXEL SIZE
14	80° X 58°	58°	0.9%	33.9
20	61° X 43°	43°	1.2%	45.8
28	45° X 31°	31°	1.7%	63.5
35	37° X 25°	25°	2.1%	78.7
50	26° X 18°	18°	2.9%	109.3
105	13° X 8°	8°	6.6%	245.9
200	6.7° X 4.5°	4.5°	11.8%	437.2
400	3.4° X 2°	2°	26.5%	983.7
500	2.7° X 1.8	1.8°	29.4%	1093.0
1000	1.3° X 0.9°	0.9°	58.9%	2186.0
1500	0.9° X 0.6°	0.6°	88.3%	3278.9
2000	0.6° X 0.45°	0.5°	117.8%	4371.9

Chart 2: APS-C Crop sensor camera field of view. The 3rd column is the vertical field of view in degrees. Column 4 is the percentage of the total sensor height that the sun covers. Column 5 is how many pixels wide the sun will be on a 12mp Nikon D300s. (Values are estimates)

Eclipse Photography « 39

The big challenge is the cost of the lens. Lenses longer than 300mm are expensive. They also require heavier tripods and specialized tripod heads. The 70-300mm lenses from Nikon, Canon, Tamron, and others are relatively affordable options. It is worth spending time at a local camera shop to try different lenses. Long focal-length lenses are a significant investment, especially for a single event.

To achieve a large eclipse image, you will need a long focal-length lens, ideally at least 400mm. A standard 70-300mm lens set to 300mm will show a small sun. At 500mm, the sun image becomes larger and covers more of the sensor area. The corona will take up a significant portion of the frame. By 1000mm, the corona will exceed the capture area on a full-frame sensor. See the picture on page thirty-seven for sun size simulations for different focal lengths.

Suggested Exposures

To photograph the partial eclipse, the camera must have a solar filter attached. If not, the intense light from the sun may damage (fry) the inside of your camera. This has happened to the author. The exposure depends on the density (darkness) of the solar filter used.

As a starting point, set the camera to ISO 100, f/8, and with the solar filter on, try an exposure of 1/4000. Make adjustments based on the histogram and highlight warning.

Turn on the highlight warning in your camera. This feature is commonly called "blinkies." This warning will help you detect if the image is overexposed or not.

Once the Baily's Beads, prominences, and corona become visible, there will only be 2.25 minutes to take bracketed shots. It will take at least eleven shots to capture the various areas of the sun's corona. The brightness varies considerably. No commercially available camera can capture the incredible dynamic range of the different portions of the delicate corona. This requires taking multiple photographs and digitally combining them afterward.

During totality, try these exposure times at ISO 100 and f/8:

1/4000, 1/2000, 1/1000, 1/250, 1/60, 1/30, 1/15, 1/4, 1/2, 1 sec, and 4 sec.

Photography Time

Set the camera to full-stop adjustments. It will reduce the time spent fiddling. As an example, the author tried the above shot sequence, adjusting the shutter speed as fast as possible.

It took thirty-three seconds to shoot the above 11 shots using 1/3-stop increments. This was without adjusting composition, focus, or anything else but the shutter speed. When the camera was set to full stop increments, it only took twenty-two seconds to step through the same shutter speed sequence.

Assuming the totality lasts less than two minutes, only four shot sequences could be made using 1/3-stop increments. Yet six shot sequences could be made when the camera was set to full stop steps. Zero time was spent looking at the back LCD to analyze highlights and the histogram.

Now add in the bare minimum time to check the highlight warning. It took sixty-three seconds to shoot and check each image using full stops. And that was without changing the composition to allow for sun movement, bumping the tripod, etc. Looking at the LCD ("chimping") consumed **half** of the totality time.

This test was done in the comfort of home under no pressure. In real world conditions, it may be possible to successfully shoot only one sequence. If you plan to capture the entire dynamic range of the totality, you must practice the sequence until you have it down cold. If you normally fumble with your camera, do not underestimate the difficulty, frustration, and stress of total eclipse photography.

Most importantly, trying to shoot this sequence allowed for zero time to simply look at the totality to enjoy the spectacle.

Avoid Last Minute Purchases

You should purchase whatever you think you'll need to photograph the eclipse today. This event will be nothing short of massive. Remember the hot toy of the year? Multiply that frenzy by a thousand. Everyone will want to try to capture their own photo.

Do not wait until the last few weeks before the eclipse to purchase cameras, lenses, filters, tripods, viewing glasses, and associated material. Consider that the totality of the eclipse will streak from coast to coast. Everyone who wants to photograph the eclipse will order at the same time. If you wait until August to buy what you need, it's conceivable that every piece of camera equipment capable of creating a total eclipse photo will be sold out in the United States. Whether this happens or not, do not wait until midsummer to make your purchases. It may be too late.

Practice

You will need to practice with your equipment. Things may go wrong that you don't anticipate. If you've never photographed a partial or total eclipse, taking quality shots is more difficult than you think. Practice shooting the sequence with a midday sun. This will tell you if you have your exposures and timing correct. Figure out what you need well in advance.

Practice photographing the full moon and stars at night. Capture the moon in full daylight. There will be six moon cycles to practice with. Astrophotography is challenging and requires practice.

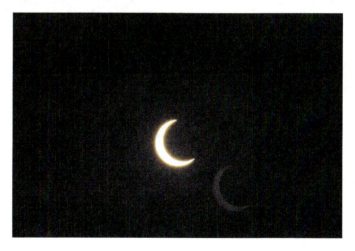

The May 20, 2012, eclipse as seen in San Diego, CA, shot with a Nikon D300s (crop sensor) with an 80-400mm lens set to ~350mm. The sun is 560 pixels wide on the 4288x2848 image.

This image is shown straight out of the camera without modification. Even with a high-quality camera and lens, photographing an eclipse is challenging. Note the haze and reflection from the overexposed sun.

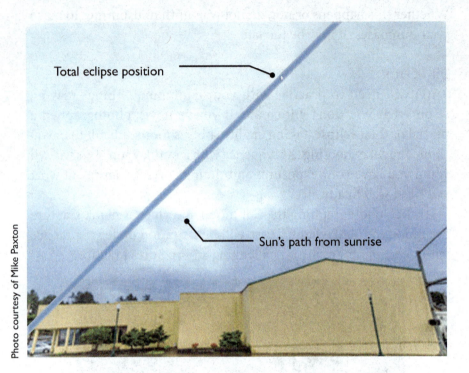

The sun will follow this path on the morning of the eclipse on August 21, 2017. The shot was taken from outside Prehistoric (1425 NW U.S. 101, Lincoln City, OR 97367).

Hopefully there will not be as many clouds on the day of the eclipse. *This image is an approximation of where the total eclipse will appear.*

☉ is the symbol for the sun and first appeared in Europe during the Renaissance.
☾ is the ancient symbol for the moon.

Viewing Locations Around Oregon

Hundreds of thousands of people will travel to and around Oregon to view the total eclipse. Most will visit a few cities and towns in western Oregon along the I-5 corridor. Due to the position of the sun when the eclipse happens, it will be difficult to capture the eclipse over sea stacks and other popular locations that are normally viewed while facing west. The sun will be in the southeast sky at the moment of totality across the state.

This section contains popular, alternative, and little-known locations to watch the eclipse. As long as there are no clouds or smoke from fires, the partial eclipse will be viewable from anywhere in Oregon.

Suggested Total Eclipse View Points

Towns and Cities

- Albany
- Baker City
- Lime
- Lincoln City
- Madras
- Mount Vernon
- Newport
- Prineville
- Salem

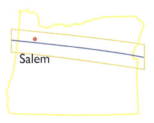

Oregon Total Eclipse Path

Unique Locations

- Beaches
- Detroit Lake
- Devils Punchbowl
- John Day Fossil Beds
- Malheur NF
- McDonald-Dunn Forest
- Ochoco NF
- Painted Hills
- Phillips Lake
- Silver Falls SP
- The Cove Palisades SP
- Umatilla NF

Albany

Elevation: 210 feet
Population: 51,583 (2013)
Main road/hwy: Hwy 19

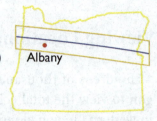

Overview

Albany is one of the larger cities in western Oregon along the path of the total eclipse. It is the seat of Linn County and is at the confluence of the Willamette River and Calapooia River.

The Albany American Legion is sponsoring an event called "Party in the Path." It will take place at the Albany Infinite Air Center. The city is still planning the event, so check their website for more details.

Getting There

The city of Albany is located west of I-5. It is located twenty-five miles south of Salem.

Totality Duration

1 minutes 52 seconds

Notes

Visit Albany's official website for updates on the total eclipse: albanyvisitors.com

Event	Time (PDT)	Altitude	Azimuth
Sunrise	6:22:00AM	0°	73°
Eclipse Start	9:05:06AM	27°	101°
Totality Start	10:17:05AM	39°	116°
Totality End	10:18:57AM	40°	117°
Eclipse End	11:37:41AM	51°	139°
Sunset	8:07:00PM	0°	287°

Baker City

Elevation: 3,451 feet
Population: 9,769
Main road/hwy: I-84

Overview

Bakery City is one of the largest communities in east Oregon that will enjoy the eclipse totality on August 21. Local Baker City author Nancy Coffelt wrote a children's book about the eclipse, titled *The Big Eclipse*. Baker County Tourism and the Chamber of Commerce are all planning for a large influx of people for this event.

The historic center of the town has been voted one of Rand McNally's six most beautiful small towns in America for three years. There are over one hundred buildings registered on the National Historic Register of Historic Places. Visit Baker's website for specific information: http://www.visitbaker.com/

Getting There

Drive north from Boise, ID, on the I-84 to reach Baker City. Alternatively, drive east on the I-84 from Portland and The Dalles.

Totality Duration

1 minutes 35 seconds

Notes

Visit the county's website for more eclipse details:
basecampbaker.com/wp/eclipse-2017

Event	Time (PDT)	Altitude	Azimuth
Sunrise	6:01:00AM	0°	73°
Eclipse Start	9:09:52AM	33°	106°
Totality Start	10:24:31AM	41°	124°
Totality End	10:26:03AM	44°	124°
Eclipse End	11:46:53AM	54°	151°
Sunset	7:46:00PM	0°	287°

Lime (Ghost Town)

Elevation: 2,234 feet
Population: Unincorporated
Main road/hwy: I-84

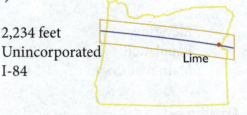

Overview

Lime is an unincorporated town along the path of totality in eastern Oregon. This is one of the unique nearly ghost towns along the eclipse path. The namesake of the town originates from the limestone in the area that supplied the cement factories in the town. The abandoned cement plant promises to be a unique and potentially spooky place to experience the eclipse from.

Getting There

Drive east for 259 miles from The Dalles to Lime on I-84.

Totality Duration

2 minutes 10 seconds

Notes

There are a handful of residents who live in Lime. Please respect their privacy and private property.

Note that the community of Lime is in the Pacific time zone. Communities a few miles south of Lime fall into the Mountain time zone.

Event	Time (PDT)	Altitude	Azimuth
Sunrise	5:59:00AM	0°	72°
Eclipse Start	9:10:04AM	32°	106°
Totality Start	10:24:49AM	44°	124°
Totality End	10:26:58AM	45°	124°
Eclipse End	11:47:54AM	54°	152°
Sunset	7:42:00PM	0°	287°

Lincoln City

Elevation: Sea Level
Population: 8,040 (2013)
Main road/hwy: US 101

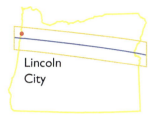

Overview

Lincoln City is one of the cities that will experience first landfall of the total eclipse. This city is a popular destination for people driving from Salem and Portland during the tourist seasons. One unique store in Lincoln City is Prehistoric (1425 NW Hwy 101), a popular shop showcasing fossils, minerals, and meteorites. Learn about the eclipse and purchase eclipse glasses here. Visit their website for more information: prehistoricoregon.com

Getting There

Drive west for fifty-eight miles from Salem on Highway 22/18 to reach Lincoln City.

Totality Duration

1 minutes 53 seconds

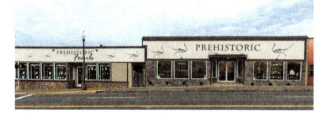

Notes

The Oregon coastline is dangerous. Never turn your back to the ocean or let children go near the water unsupervised. Sneaker waves have claimed many lives over the year.

Event	Time (PDT)	Altitude	Azimuth
Sunrise	6:25:00AM	0°	72°
Eclipse Start	9:04:41AM	27°	100°
Totality Start	10:16:07AM	38°	115°
Totality End	10:18:01AM	39°	116°
Eclipse End	11:36:11AM	50°	138°
Sunset	8:11:00PM	0°	287°

Above photo courtesy of Mike Paxton

Madras

Elevation: 2,241 feet
Population: 6,363 (2013)
Main road/hwy: US 97

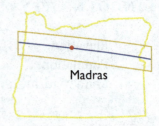

Overview

The town of Madras is one of the larger communities in central Oregon that is nearly on the centerline of the total eclipse. The Lowell Observatory from Flagstaff, AZ, is working with the community to put on the Lowell Observatory's 2017 Solar Eclipse Experience at the Madras High School football field. Check their website for more information: www.lowellsolareclipse.com. There is also another event in Madras called the Oregon Solarfest from August 17-22. This event is partnered with NASA. Visit their website for more information about attending: www.oregonsolarfest.com.

Getting There

Drive north from Redmond for twenty-six miles on US 97 to reach the town of Madras.

Totality Duration

2 minutes 2 seconds

Notes

The community expects over 100,000 visitors for the various events around town. Plan accordingly.

Event	Time (PDT)	Altitude	Azimuth
Sunrise	6:14:00AM	0°	73°
Eclipse Start	9:06:43AM	29°	102°
Totality Start	10:19:36AM	41°	119°
Totality End	10:21:38AM	41°	119°
Eclipse End	11:41:05AM	52°	143°
Sunset	7:59:00PM	0°	287°

Mount Vernon

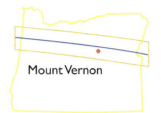

Elevation: 2,865 feet
Population: 513 (2013)
Main road/hwy: US 26

Overview

The small town of Mount Vernon is along the path of the eclipse totality. The town was named after a black stallion owned by a settler by the name of David W. Jenkins. There is a hot springs resort approximately two miles north of town. The town is along the John Day River that will be a cool place to watch the total eclipse from, as summertime temperatures can rise to nearly ninety degrees in August.

Getting There

Starting at Prineville, drive east 108 miles on US 26 to reach the town of Mount Vernon.

Totality Duration

2 minutes 0 seconds

Notes

Mount Vernon is one of several communities along US 26 from which to watch the total eclipse. If one is too crowded, visit another. Plan for heavy traffic along US 26.

Event	Time (PDT)	Altitude	Azimuth
Sunrise	6:07:00AM	0°	72°
Eclipse Start	9:08:20AM	31°	104°
Totality Start	10:22:16AM	43°	121°
Totality End	10:24:17AM	43°	122°
Eclipse End	11:44:39AM	53°	148°
Sunset	7:51:00PM	0°	287°

Newport

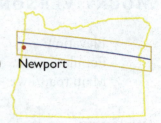

Elevation: Sea level
Population: 10,117 (2013)
Main road/hwy: US 101

Overview

Newport is one of the first contact points of the moon's shadow during the total eclipse. Located south of the centerline, this beach community will be loaded with visitors on eclipse day. Finding lodging will be difficult, as visitors have known about this event for some time. The coast of Oregon is spectacular with sea stacks (rocks in the ocean), making for an energetic setting for the eclipse. As the sun will be in the southeast sky, it will not be possible to photograph sea stacks and the eclipse without a boat.

Getting There

Drive west from Corvallis for forty-eight miles on US 20 to reach the city of Newport.

Totality Duration

1 minutes 45 seconds

Notes

Note the Oregon coastline is dangerous. Never turn your back to the ocean. Sneaker waves have claimed many lives over the year. Watch your children and beware of floating logs.

Event	Time (PDT)	Altitude	Azimuth
Sunrise	6:26:00AM	0°	72°
Eclipse Start	9:04:21AM	26°	100°
Totality Start	10:15:55AM	39°	115°
Totality End	10:17:41AM	39°	115°
Eclipse End	11:36:04AM	50°	162°
Sunset	8:11:00AM	0°	287°

Prineville
Big Summit Prairie
Crook County

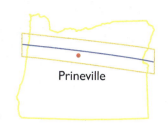

Elevation: 2,868 feet
Population: 9,223 (2013)
Main road/hwy: Hwy 7

Overview

The city of Prineville is located west of the Ochoco National Forest. It is the county seat of Crook County. It was chosen as the site of one of the major eclipse gatherings in the state of Oregon. The festival will be held at Big Summit Prairie, a private 55,000-acre ranch with camping areas and a fifty-two-acre lake.

Getting There

Starting at Redmond, drive twenty miles east on Highway 126 to reach the city of Prineville.

Totality Duration

1 minutes 14 seconds

Notes

Check the Oregon Eclipse website for more details about the Total Eclipse Gathering (Aug 17-23): oregoneclipse2017.com

Event	Time (PDT)	Altitude	Azimuth
Sunrise	6:14:00AM	0°	72°
Eclipse Start	9:06:41AM	29°	103°
Totality Start	10:20:13AM	41°	119°
Totality End	10:21:24AM	42°	119°
Eclipse End	11:41:35AM	52°	144°
Sunset	7:57:00PM	0°	287°

Salem

Elevation: 153 feet
Population: 160,614 (2013)
Main road/hwy: I-5

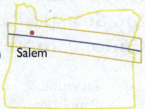

Overview

Salem is one of several state capitals along the line of the 2017 total eclipse. As this city is large, they will be more able to handle the massive influx of visitors compared to other small communities in Oregon. Nevertheless, expect massive crowds from the greater Portland metro area to drive to Salem to watch the eclipse. There are a number of interesting viewing locations. If you find yourself stuck in unmoving traffic, simply watch the eclipse from where you are. It is high enough in the sky that only a tall building will block your view.

Getting There

Capital of Oregon. Drive south from Portland on I-5 for forty-six miles.

Totality Duration

1 minutes 51 seconds

Notes

Although Salem is a large city, the author was unable to find any hotels on Hotels.com from August 18 to August 22. Consider using web rental services such as Airbnb.com to find lodging for this event.

Event	Time (PDT)	Altitude	Azimuth
Sunrise	6:21:00AM	0°	73°
Eclipse Start	9:05:26AM	27°	101°
Totality Start	10:17:21AM	39°	116°
Totality End	10:19:15AM	40°	117°
Eclipse End	11:37:50AM	50°	140°
Sunset	8:07:00PM	0°	287°

BEACHES

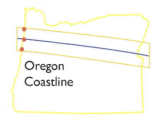

Elevation: Sea level
Main road/hwy: US 101

OVERVIEW

Oregon is unique in that it is the only state that the total eclipse will come into first contact with when it makes landfall.

Wakonda Beach is outside the path of totality. Waldport is inside the south edge of the path. The Sand Lake Recreational Area is on the northern edge of totality. Fogarty Creek State Recreation Area is on the centerline.

Areas first touched by eclipse: (south) Yaquina Head Outstanding Natural Area, Bureau of Land Management to Depoe Bay (north) are the first contact zones of the total eclipse.

GETTING THERE

Drive along US 101 to reach any of these locations.

TOTALITY DURATION

Varies depending on location

NOTES

Note the Oregon coastline is dangerous. Never turn your back to the ocean. Sneaker waves have claimed many lives over the year. Watch your children and beware of floating logs.

Chart times for Depoe Bay, OR, near the center of totality.

Event	Time (PDT)	Altitude	Azimuth
Sunrise	6:25:00AM	0°	72°
Eclipse Start	9:04:30AM	26°	100°
Totality Start	10:15:55AM	39°	115°
Totality End	10:17:53AM	39°	115°
Eclipse End	11:36:05AM	50°	138°
Sunset	8:11:00PM	0°	287°

Detroit Lake State Recreation Area

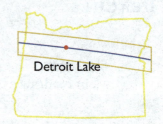

Elevation: 1,558 feet
Main road/hwy: Hwy 22

Overview

Detroit Lake near the community of Detroit is named for the large number of people who came from Detroit, Michigan, in the 1890s. The reservoir was created by the Detroit Dam on the North Santiam River. The lake is 5.5 square miles and is managed by Oregon State Parks and Recreation. Although regular camping has been sold out since 2016, it may be worth checking the Parks and Recreation website for more opportunities.

Getting There

Drive from Salem to the lake on Highway 22.

Totality Duration

2 minutes 3 seconds

Notes

There are multiple towns and cities along Highway 22. Expect significant traffic on the day of the eclipse.

Event	Time (PDT)	Altitude	Azimuth
Sunrise	6:18:00AM	0°	72°
Eclipse Start	9:05:57AM	28°	102°
Totality Start	10:18:18AM	40°	117°
Totality End	10:20:20AM	40°	118°
Eclipse End	11:39:20AM	51°	141°
Sunset	8:03:00PM	0°	287°

Devils Punchbowl State Natural Area

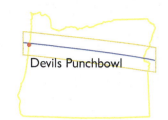

Elevation:	Sea Level
Main road/hwy:	US 101

Overview

Devils Punchbowl is one of the first locations that the total eclipse will make landfall on the continental United States. The ocean cut a hole in the coastal rock and, over time, created a bridge. This allows the seawater to flow in and out of the natural formation. The theory is that the bowl was created when two adjoining sea caves collapsed.

Getting There

Devils Punch Bowl is located approximately five miles south of Depoe Bay in the community of Otter Rock, 1/4 mile west of US 101.

Totality Duration

1 minutes 57 seconds

Notes

Note that the water in the bowl is extremely dangerous. As with all Oregon coastline, be mindful of yourself and any young people with you. The rocky water is dangerous and has claimed countless lives over the years. Unexpected sneaker waves seem to appear out of nowhere. Do not underestimate the coastline.

Event	Time (PDT)	Altitude	Azimuth
Sunrise	6:26:00AM	0°	72°
Eclipse Start	9:04:26AM	26°	100°
Totality Start	10:15:54AM	39°	115°
Totality End	10:17:49AM	39°	115°
Eclipse End	11:36:04AM	50°	138°
Sunset	8:11:00PM	0°	287°

John Day Fossil Beds NM

Elevation: Various
Main road/hwy: US 26

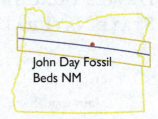

Overview

John Day Fossil Beds National Monument is slightly south of the total eclipse centerline. This monument has several different districts to visit. They include the Painted Hills discussed elsewhere in this book. The fossil beds are a veritable history book of ancient life in central Oregon. The landscape is unique and will make fascinating photographs with the otherworldly total eclipse light.

Getting There

Drive east for ninety-five miles on US 26 until you reach Highway 19. Turn north into the John Day Fossil Beds.

Totality Duration

2 minutes 5 seconds

Notes

Visit the monument's official website for more information about access and visitation:
www.nps.gov/joda

Event	Time (PDT)	Altitude	Azimuth
Sunrise	6:08:00AM	0°	73°
Eclipse Start	9:07:57AM	30°	104°
Totality Start	10:21:34AM	42°	121°
Totality End	10:23:39AM	43°	121°
Eclipse End	11:43:43AM	53°	146°
Sunset	7:52:00PM	0°	287°

Malheur National Forest

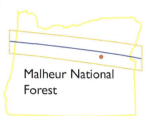

Elevation: Various
Main road/hwy: US 395

Overview

Covering 1.7 million acres of forested land, the northern half of the Malheur National Forest will experience the August 21 total eclipse. Bisected by Highway 395, this forest covers vast tracts of land and topography in eastern Oregon. The towns of Seneca, Silvies, and Trout Creek are surrounded by the Malheur National Forest, but they are not actually on forest service land. These different towns provide access to much of the center of the Malheur National Forest.

Getting There

Travel east from Redmond on Route 126 to Prineville. From there, continue east on US 26 until you reach the town of John Day. Turn south on US 395 to reach the towns mentioned above to access the forest.

Totality Duration

Varies depending on location

Notes

Visit this webpage for more information about the eastern Oregon total eclipse: easternoreclipse.blogspot.com

Chart times for Seneca, OR. Times vary depending on location.

Event	Time (PDT)	Altitude	Azimuth
Sunrise	6:07:00AM	0°	72°
Eclipse Start	9:08:14AM	31°	104°
Totality Start	10:22:54AM	43°	121°
Totality End	10:23:48AM	43°	122°
Eclipse End	11:44:56AM	54°	148°
Sunset	7:49:00PM	0°	287°

McDonald-Dunn Forest

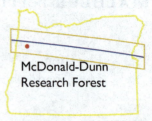

Elevation: 748 feet
Main road/hwy: Hwy 99W

Overview

Located near Corvallis, this forest is managed by the Oregon State University College of Forestry. The university conducts long-term surveys about the growth of trees and tests forest management techniques. To experience the total eclipse in a natural surrounding, residents of Corvallis, Albany, and surrounding cities only have to drive a few miles to reach this unique and historic 11,250 acres of predominantly forested land.

Getting There

Driving directions from Corvallis can be found on this webpage: http://cf.forestry.oregonstate.edu/how-find-us

Totality Duration

Varies depending on location

Notes

More history about the importance and struggles about how this forest research area came to be can be found here:
cf.forestry.oregonstate.edu/acquisition-mcdonald-dunn-forests

Chart times for the center of the forest.

Event	Time (PDT)	Altitude	Azimuth
Sunrise	6:23:00AM	0°	72°
Eclipse Start	9:04:55AM	27°	100°
Totality Start	10:16:48AM	39°	116°
Totality End	10:18:39AM	40°	116°
Eclipse End	11:37:17AM	51°	139°
Sunset	8:08:00PM	0°	287°

Ochoco National Forest

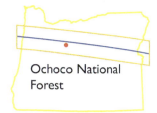

Elevation: Various
Main road/hwy: US 26

Overview

Nearly all the Ochoco National Forest will be covered by the totality during the eclipse on August 21. This forest of 850,000 acres is filled with dense pine forests, peculiar geologic formations, and high desert terrain. Nearly anywhere in this national forest will be a good viewing location, assuming there are no forest fires in the region. The nearby town of Prineville (west of the Ochoco National Forest) is hosting a major music and entertainment festival during the days leading up to the total eclipse.

Getting There

Travel east from Redmond on Route 126 to Prineville. From there, continue east on US 26 until you reach the national forest boundary.

Totality Duration

Varies depending on location

Notes

Much of the forest has service roads. Check ahead before planning on driving these roads, as conditions may change.

Chart times for Bandit Springs, OR.

Event	Time (PDT)	Altitude	Azimuth
Sunrise	6:12:00AM	0°	72°
Eclipse Start	9:07:13AM	30°	103°
Totality Start	10:20:33AM	42°	120°
Totality End	10:22:29AM	42°	120°
Eclipse End	11:42:22AM	52°	145°
Sunset	7:56:00PM	0°	287°

Painted Hills

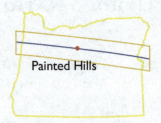

Elevation: 2,114 feet
Main road/hwy: US 26

Overview

The Painted Hills is one of three operations units that make up the John Day Fossil Beds National Monument in Oregon. The location is named for the delicate layers of black, gold, red, and yellow soils. Named one of the Seven Wonders of Oregon by one blogger, this location promises to be a visual feast during the eclipse. As this location is directly on the centerline of totality, the light will be surreal as the totality nears. This may be one of the most unique spots to create unforgettable photographs from. Stop by the town of Mitchell before visiting the Painted Hills for updates on conditions.

Getting There

Drive northeast forty-three miles on US 26 from Prineville. Turn left on Burnt Ranch Rd. that becomes Bridge Creek Rd. Turn left on Bear Creek Rd. and stop at the Painted Hills Viewpoint.

Totality Duration

2 minutes 5 seconds

Notes

Visit the town of Mitchell webpage for more information: mitchelloregon.us/painted-hills

Event	Time (PDT)	Altitude	Azimuth
Sunrise	6:11:00AM	0°	73°
Eclipse Start	9:07:30AM	30°	103°
Totality Start	10:20:46AM	42°	120°
Totality End	10:22:50AM	42°	120°
Eclipse End	11:42:36AM	52°	145°
Sunset	7:56:00PM	0°	287°

Phillips Lake

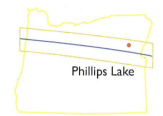

Elevation: 4,068 feet
Main road/hwy: Hwy 7

Overview
Located near the town of McEwen, Phillips Lake promises to be an excellent spot to watch the total eclipse from. This reservoir has a wide range of activities for visitors, such as hiking, water sports, camping, and fishing. The 2,235-acre lake offers eclipse chasers the chance to watch the totality over water. This reduces the amount of atmospheric distortion, improving the view and cooling the air down. If possible, watch the eclipse from the west side of the reservoir for the best atmospheric cooling.

Getting There
Starting from Baker City, drive south on Highway 7. Remain on Highway 7 when the road intersects Highway 245. Continue west until you reach the lake.

Totality Duration
1 minutes 52 seconds

Notes
Be aware that camping reservations may be difficult to come by. Contact the Wallowa-Whitman National Forest for more information.

Event	Time (PDT)	Altitude	Azimuth
Sunrise	6:02:00AM	0°	72°
Eclipse Start	9:09:35AM	32°	106°
Totality Start	10:24:00AM	44°	123°
Totality End	10:25:54AM	44°	124°
Eclipse End	11:46:32AM	54°	150°
Sunset	7:47:00PM	0°	287°

Silver Falls State Park

Elevation: 1,590 feet
Main road/hwy: Hwy 7

Overview

Located twenty-five miles east of Salem, Silver Falls State Park is one of the state parks that are nearly on the centerline of the eclipse. The centerpiece and namesake of the park is the beautiful waterfall that creates a 177-foot curtain of water. It dives and splashes throughout the year. This park is referred to as the crown jewel of the Oregon State Park system. There are multiple trails of varying difficulty for families to enjoy before and after the total eclipse. The park began receiving reservations to camp over a year ago. The park reports there may be other campsites available.

Getting There

Drive southeast along Highway 22 for ten miles. Take exit 9 onto Highway 214. Drive another fourteen miles until you reach the park.

Totality Duration

1 minutes 57 seconds

Notes

Find more information about Silver Falls State Park at the Oregon State Park website: oregonstateparks.org.

Event	Time (PDT)	Altitude	Azimuth
Sunrise	6:20:00AM	0°	71°
Eclipse Start	9:05:41AM	28°	101°
Totality Start	10:17:46AM	40°	117°
Totality End	10:19:44AM	40°	117°
Eclipse End	11:38:29AM	51°	141°
Sunset	8:06:00PM	0°	287°

The Cove Palisades State Park

Elevation: 1,948 feet
Distance: Drive to SP
Main road/hwy: US 97

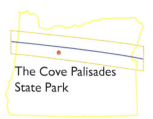
The Cove Palisades State Park

Overview

The Cove Palisades State Park will be a fascinating location to view the total eclipse from. Located south of the centerline, most of the shoreline runs north-south. This means that there are countless locations to view the total eclipse over the water. The Cove Palisades Resort will be a central location to meet for groups driving to this state park. Note that like all other state parks, standard campsites sold out last year. Contact the state park for updated information.

Getting There

Drive north on US 97 from Redmond. Turn onto SW Culver Hwy to reach this state park.

Totality Duration

1 minutes 55 seconds

Notes

Visit the Cove Palisades' The Cover Rattle page for updates and more information about the total eclipse:

covepalisades.wordpress.com/2017-total-solar-eclipse-information/

Event	Time (PDT)	Altitude	Azimuth
Sunrise	6:15:00AM	0	72°
Eclipse Start	9:06:30AM	29°	102°
Totality Start	10:19:24AM	41°	119°
Totality End	10:21:22AM	41°	119°
Eclipse End	11:40:50AM	52°	143°
Sunset	7:59:00PM	0°	287°

Umatilla National Forest

Elevation: Various
Main road/hwy: US 26/395

Umatilla National Forest

Overview

The Umatilla National Forest covers 2,199 square miles (1.4 million acres) and stretches across a large swath of eastern Oregon. The total eclipse path will cross over the southern section of this forest area located in the Blue Mountains of northeast Oregon and southeast Washington.

Getting There

There are multiple access points to reach this forest. From Prairie City, drive east on US 26 until you reach the forest boundary. Highway 395 roughly follows the western edge of the forest, north of Mount Vernon and south of Ukiah.

Totality Duration

Varies depending on location

Notes

Find out more information about the total eclipse in the Umatilla National Forest by visiting their total eclipse webpage: www.fs.usda.gov/detail/umatilla/home/?cid=fseprd536830

Chart times for Galena, OR. Times vary depending on location.

Event	Time (PDT)	Altitude	Azimuth
Sunrise	6:05:00AM	0°	72°
Eclipse Start	9:08:52AM	31°	105°
Totality Start	10:22:52AM	43°	122°
Totality End	10:24:50AM	43°	123°
Eclipse End	11:45:09AM	53°	148°
Sunset	7:50:00PM	0°	287°

Remember the Oregon Total Eclipse
August 21, 2017

Who was I with? _____

_____

_____

_____

_____

What did I see? _____

_____

_____

_____

_____

_____

_____

_____

_____

What did I feel? _____

_____

_____

_____

_____

_____

_____

_____

_____

What did the people with me think? _____

Where did I stay? _____

Enjoy other Sastrugi Press titles

2017 Total Eclipse State Series by Aaron Linsdau

Sastrugi Press has published several state-specific guides to the 2017 total eclipse crossing over the United States. Check the Sastrugi Press website for the various state eclipse books: www.sastrugipress.com/eclipse/

Antarctic Tears by Aaron Linsdau

What would make someone give up a high-paying career to ski alone across Antarctica? This inspirational true story will make readers both cheer and cry. Fighting skin-freezing temperatures, infections, and emotional breakdown, Aaron Linsdau exposes the harsh realities of the world's largest wilderness. Discover what drives someone to the brink of destruction while pursuing a dream.

Adventure One by Aaron Linsdau and Terry Williams, M.D.

What does it take to conceptualize, plan, and enjoy your first expedition? This inspirational book contains hard-won knowledge from both authors about their experiences on expeditions around the world. The information provided in this book is useful whether you plan to climb a high peak, cross a polar plateau, or set out on a never before attempted new trek. (*Available Summer 2017*)

Lost at Windy Corner by Aaron Linsdau

Climbing Denali is a treacherous affair. Avalanches, blinding blizzards, and crevasses have killed experienced teams. What happens when someone decides to climb the mountain solo? In this dramatic story, Aaron describes the choices made and the lessons that were learned as a result. This is more than an adventure story. It teaches defining success on your own terms in business and life. The messages will stay with you long after the end of the book.

Visit Sastrugi Press on the web at www.sastrugipress.com to purchase the above titles in bulk. They are also available from your local bookstore or online retailers in print, e-book, or audiobook form.

Thank you for choosing Sastrugi Press.
"Turn the Page Loose"

About Aaron Linsdau

Aaron Linsdau is a polar explorer and motivational speaker. He energizes audiences with life and business lessons that stick. He delivers a message of courage by building grit and maintaining a positive attitude. Aaron teaches audiences how to eat two sticks of butter a day to achieve their goal. He shares how to build resilience to deal with constant pressure and adrenaline overload.

He holds the world record for the longest expedition in days from Hercules Inlet to the South Pole. Aaron is the second only American to complete the trip alone.

This solo expedition is more difficult than climbing Mount Everest with a team. Being alone dramatically increases the challenge. Aaron uses emotionally stirring stories to show how to overcome obstacles, impossible challenges, and unimaginable conditions. He relates these stories to business challenges and shows how the common person can achieve uncommon results.

Aaron collaborates with organizations to deliver the right message for the audience. He relates his experiences to business realities. Aaron loves inspiring audiences. Book Aaron for your next event today.

"Never Give Up"
Grit • Courage • Attitude • Perseverance • Resilience

Learn more about Aaron Linsdau at:
www.aaronlinsdau.com or www.ncexped.com.

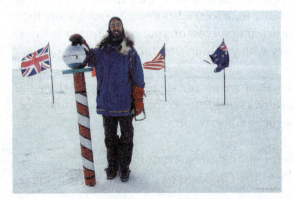

Aaron at the South Pole after 82 days alone in Antarctica.

Smartphone link

CPSIA information can be obtained
at www.ICGtesting.com
Printed in the USA
FSOW04n1121200717
36550FS